About Demos

Demos is a greenhouse for new ideas which can improve the quality of our lives. As an independent think tank, we aim to create an open resource of knowledge and learning that operates beyond traditional party politics.

We connect researchers, thinkers and practitioners to an international network of people changing politics. Our ideas regularly influence government policy, but we also work with companies, NGOs, colleges and professional bodies.

Demos knowledge is organised around five themes, which combine to create new perspectives. The themes are democracy, learning, enterprise, quality of life and global change.

But we also understand that thinking by itself is not enough. Demos has helped to initiate a number of practical projects which are delivering real social benefit through the redesign of public services.

We bring together people from a wide range of backgrounds to cross-fertilise ideas and experience. By working with Demos, our partners develop a sharper insight into the way ideas shape society. For Demos, the process is as important as the final product.

www.demos.co.uk

D0263167

First published in 2004
© Demos
Some rights reserved – see copyright licence for details

ISBN 1 84180 130 5
Typeset by Land & Unwin, Bugbrooke
Printed by HenDI Systems, London

For further information and
subscription details please contact:

Demos
Magdalen House
136 Tooley Street
London SE1 2TU

telephone: 0845 458 5949
email: hello@demos.co.uk
web: www.demos.co.uk

See-through Science

Why public engagement needs to move upstream

James Wilsdon
Rebecca Willis

Foreword by Barbara Young

DEM⊙S

DEM⊙S

Contents

Acknowledgements

This pamphlet started life as a paper for the July 2003 Progressive Governance summit, which was subsequently published in Tony Giddens' edited collection *The Progressive Manifesto* (Polity Press 2003). We are grateful to everyone who contributed to that initial version of the argument.

The Environment Agency has generously supported us in producing a fuller account of the case for 'upstream' public engagement. Many thanks to those at the Agency who provided helpful input: John Colvin, Michael Depledge, Phil Irving, Jimi Irwin, Peter Madden and Ronan Palmer.

Thanks also to Penny Egan, Susie Harries, Simon Lock and the steering group of the RSA's Forum for Technology, Citizens and the Market for their involvement in the project, and for generously offering to host the launch event.

We are grateful to those we interviewed or who informed our thinking during the research process, including: Rob Doubleday, Sue Mayer, Monica Winstanley, George Smith, Nick Pidgeon, Jim Thomas, Andy Stirling, Kathy Sykes and Gary Kass. At Demos, Tom Bentley gave us his usual mix of valuable advice and support, Briony Greenhill provided excellent research assistance, and Eddie Gibb, Paul Miller and Claire Ghoussoub made important contributions at key stages. Thanks also to Guy Thompson and Jennie Oldham at Green Alliance.

Above all, we owe an enormous intellectual debt to our friends and colleagues at Lancaster University: Brian Wynne, Robin Grove-White, Phil Macnaghten and Matthew Kearnes. For more than a decade, the 'Lancaster school' has pioneered a way of thinking about science, technology and society that has inspired and informed the argument in these pages. They and others will spot where we have drawn from their ideas. Any errors or omissions remain entirely our own.

James Wilsdon
Rebecca Willis
August 2004

The authors

James Wilsdon is Head of Strategy at Demos, where he leads research on science, technology and sustainable development. His recent publications include: *Masters of the Universe: science, politics and the new space race* (with Melissa Mean, Demos, 2004); *The Adaptive State: strategies for personalising the public realm* (ed. with Tom Bentley, Demos, 2003); and *Digital Futures: living in a networked world* (ed., Earthscan 2001). He is Chair of People and Planet, an Associate Director of Forum for the Future, and an Honorary Research Fellow of the Institute for Environment, Philosophy and Public Policy at Lancaster University; (james.wilsdon@demos.co.uk).

Rebecca Willis is Associate Director of Green Alliance and Vice-Chair of the UK's Sustainable Development Commission. Previously she was a policy adviser at the European Parliament. Her recent publications include: *Next Steps for Energy Taxation* (with Paul Ekins et al., PSI/Green Alliance, 2002); *Precaution in Practice: how the precautionary principle is used by government, business and NGOs* (with Jennie Oldham, Green Alliance, 2002); *Steps into Uncertainty: handling risk and uncertainty in environmental policy making* (with Beatrice Rose, Green Alliance/ESRC, 2000); (rwillis@green-alliance.org.uk).

Foreword

Barbara Young

Technological innovation is vital both to the economy and to the environment. Innovation drives much of our economic growth. And new technologies can deliver solutions to environmental problems.

For example, new renewable energy technologies will help us to tackle climate change. Devices to save water will protect river catchments and sensitive ecosystems. Biological waste treatments and soil remediation can reduce pressure on landfill sites. And hydrogen fuel cells will cut air pollution.

New technology can also help us do our job more effectively. Data from environmental sensors can now be relayed via mobile phones and satellites to give us a real-time picture of water quality in our rivers. This will allow us to respond quickly to pollution incidents. New communications technologies allow us to warn people about impending floods.

As a regulatory Agency, we know that, designed in the right way, regulation can stimulate innovation. We see this as core to our task of delivering modern regulation.

The Government's Innovation Review has set out a number of new policy initiatives to promote environmental innovation. The Agency is working with DTI and Defra on research into how regulation and product policy can stimulate innovations to improve resource efficiency. We are also part of the DTI Environmental Innovations Advisory Group, which aims to identify and resolve any regulatory barriers to environmental innovation. We welcome early engagement with the developers of new technologies, to help us plan for any changes we will have to make to the way we work.

As well as bringing opportunities, the very rapid development and dissemination of technology also brings challenges. We regulate many facilities where new technologies are developed and applied. In some cases we can adapt existing approaches but, in others, new ways of regulating will be required. We must be flexible and keep pace with technological developments in order to protect the environment.

Broader societal acceptance of new technologies, especially where they are novel and raise concerns, requires open dialogue throughout the development process. If opportunities are to be realised, then engagement and dialogue must take place at the right time and involve the right people. Information must be made available in a format that can be understood easily. All sides in the debate have responsibilities: scientists must be willing to answer questions openly and honestly, industry must engage early and widely, and pressure groups and the media must be responsible in their use of science.

I firmly believe that the economy, society and the environment will benefit from more public engagement in research and development. Opening up the world of research and encouraging scientists to acknowledge the broader social and economic context within which their research will be applied should deliver more useful scientific outputs. But these must address the issues that matter to those affected by the technologies.

It is my hope that this progressive agenda will deliver scientific outputs that help us to understand better the range of technical and social issues that can impact on the environment. I also hope that public trust in government and its agencies will be enhanced if the role of scientific information in the decision-making process becomes clearer. By working together we can build public confidence in science and reap the benefits of well-directed innovation. We seek opportunities to work with others towards this goal and welcome this pamphlet as a helpful contribution to the debate.

Barbara Young is Chief Executive of the Environment Agency

1. The stage is set

On 29 July 2004, a mile down the road from London's West End, the curtain rose on a unique theatrical experiment. Its theme was nanotechnology, the science of small things. The venue was the Royal Society, the headquarters of Britain's scientific elite. The immediate audience was a group of journalists, but the performance was then relayed to a larger public: policy-makers, scientists, business leaders and campaigners, all eagerly waiting to interpret its meaning.

We live in a political culture that is steeped in science. When faced with dilemmas over food safety or phone masts, climate change or child vaccination, the first response of politicians and regulators is to seek refuge in 'sound science' and the advisors who produce it. These experts and the panels and committees they inhabit are vital to the smooth running of our political system. But how do they make their advice credible to a sceptical public? What techniques of legitimacy do they use?

One answer is that they act. According to the sociologist Stephen Hilgartner, expert advice is a form of drama. Hilgartner uses the metaphor of performance to explain how scientific advisors speak with authority on the public stage. Describing a series of reports on diet and health issued by the US National Academy of Sciences in the 1980s, he draws our attention to the theatrical dynamics at work in the production, unveiling and dissemination of expert opinion. 'Reports and recommendations are performances; advisors are performers who display their work before audiences.'[1] Before a performance starts, there are months of rehearsal and negotiation between actors. Once it is under way, a division is maintained between the back-stage, where scaffolding, costumes and props are hidden from view, and the front-stage, which is open to public scrutiny.

The Royal Society's production on 29 July surprised the critics. Its year-long inquiry into the health, environmental, ethical and social implications of nanotechnology had resulted in a report of unusual quality.[2] A few predictable voices remained unconvinced, but the majority agreed that it was a sparkling performance. Several aspects were striking:

It relied on an ensemble cast. On the Royal Society's working group, alongside the usual principals – eminent professors of physics, medicine, chemistry and engineering, the head of a Cambridge college and a senior industrialist – were some unexpected supporting players – an environmentalist, a social scientist and a consumer champion. For inquiries of this nature, such voices are often called to give evidence, but for them to sit as equals alongside 'real' scientists is rare.

It was imaginatively staged. In any performance, the stage management determines which elements are visible to the audience and which remain invisible. Typically, the work of scientific advisors takes place out of sight. Debates rage and arguments are resolved in private, long before the public is presented with a consensus view. This inquiry tried hard to be more open. Aspects of its performance were still carefully rehearsed, but there was also room for improvisation. It consulted widely, ran workshops with stakeholders, and published evidence on its website.

It was deliberately avant-garde. Anyone familiar with the Royal Society's oeuvre could spot instantly that here the style had changed. The tone was unusually precautionary. Social and ethical issues received prominent billing. Uncertainty and dialogue were recurring motifs. One actor was particularly well placed to observe these changes. Professor Nick Pidgeon sat on the nanotechnology working group, but also played a leading role in an earlier Royal Society production: its 1992 report on risk.[3] Then, the mere suggestion that risk is socially constructed – a heretical notion to many natural scientists – led to the report being downgraded and released without the society's full endorsement. Twelve years on, the mood was very different. 'A new understanding of science and society is spreading through the work of the Royal Society,' observes Pidgeon. 'These perspectives are finally being mainstreamed.'[4]

The nanotechnology report represents a change in the scientific community's approach to the risks, uncertainties and wider social implications of new and emerging technologies. In many ways, it redefines the genre. But to fully appreciate its significance, we need to locate it in the wider context of relations between science and society.

Three phases of public engagement

Historically, the authority and legitimacy of science as a public good rested on a perceived division of labour between academia, commerce and politics. Academic scientists carried out basic research in laboratories, motivated purely by the spirit of inquiry. The results of their endeavours were then applied as technology, but clear dividing lines separated the worlds of science and business. Politics only entered the fray in order to regulate the market, manage risks or set standards.

Such divisions were never as clear cut or straightforward, but they have now entirely broken down. Discrete categories of basic and applied research no longer hold in a world where the production and uses of science are intertwined and embedded in dense relationships with business and politics. This blurring of boundaries has contributed to a climate where science no longer has an automatic claim to authority and respect. As the controversies over BSE, genetically-modified (GM) crops and foods, and now nanotechnology illustrate, people are questioning scientists more and trusting them less. There is particular wariness towards scientists working in industry and government, and a suspicion of private ownership of scientific knowledge. Drawing on extensive polling data, Ben Page of MORI sums up the current state of public opinion: 'Blind faith in the men in white coats has gone and isn't coming back.'[5]

This is not a surprise. As we move towards knowledge societies that rely on innovation to drive economic growth, science and technology are likely to become increasingly contested sites of public debate. As Sheila Jasanoff notes, such 'far-reaching alterations in the nature and distribution of resources and the roles of science, industry and the state could hardly occur without wrenching political conflicts'.[6] Nanotechnology is only one of several areas where the pace of innovation is accelerating. Others such as genomics, neuroscience, pervasive computing and artificial intelligence are giving rise to distinct sets of ethical and social dilemmas.

The response of the science establishment to these fluctuating and unpredictable cycles of public and media response has been to reach out and experiment with new forms of public engagement. They are not alone in this. Government repeatedly goes through the same cycle. Confronted with public ambivalence or outright hostility towards different forms of technological, social or political innovation – whether stem cell research, reform of the NHS or war in Iraq – the standard political response is a promise to listen harder. Ours is an era of 'big conversations', of government by focus group and MORI poll.

But as New Labour has discovered to its cost, talk of engagement can backfire unless it has a demonstrable impact. Those whose engagement is being sought need to know that their participation will affect the policies and processes under discussion. They want assurance that trajectories of change and innovation will take meaningful account of their views.

This same principle applies to public engagement in science. For the past 20 years, in response to a perceived 'crisis of trust', scientists have been slowly inching their way towards involving the public in their work. They looked first to education as the answer, and more recently to processes of dialogue and participation. But these efforts, while admirable, have not yet proved sufficient. Our argument in this pamphlet is that we are on the cusp of a new phase, in which public engagement moves upstream.

Phase 1: Public understanding of science (PUS)

The initial response of scientists to growing levels of public detachment and mistrust was to embark on a mission to inform. Attempts to gauge levels of public understanding date back to the early 1970s, when annual surveys carried out by the US National Science Foundation regularly uncovered gaps in people's knowledge of scientific facts (for example, whether the earth goes round the sun or vice versa).[7] Walter Bodmer's 1985 report for the Royal Society placed PUS firmly on the UK agenda, and proclaimed 'It is clearly a part of each scientist's professional responsibility to promote the public understanding of science.'[8] The Bodmer report gave birth to a clutch of initiatives designed to tackle the blight of public ignorance, including COPUS, the Committee on the Public Understanding of Science.

Phase 2: From deficit to dialogue

For more than a decade, the language and methods of PUS oozed across the face of UK science policy. But instead of lubricating understanding, scientists gradually discovered that PUS was clogging the cracks and pores which might have allowed genuine dialogue to breathe. Implicit within PUS was a set of questionable assumptions about science, the public and the nature of understanding. It relied on a 'deficit model' of the public as ignorant and science as unchanging and universally comprehensible. Partly as a result of PUS's failings, relations between science and society festered throughout the 1990s, and an occasional rash of blisters erupted (the BSE crisis, GM crops, mobile phones, MMR). It wasn't until 2000 that PUS was washed away, when an influential House of Lords report detected 'a new mood for dialogue'.[9] Out went PUS, which even the government's Chief Scientific Adviser now acknowledged was 'a rather backward-looking vision'.[10] In came the new language of 'science and society' and a fresh impetus towards dialogue and engagement.

Phase 3: Moving engagement upstream

The House of Lords report detected 'a new humility on the part of science in the face of public attitudes, and a new assertiveness on the part of the public'.[11] And in the four years since it was published, there has been a perceptible change. Consultation papers, focus groups, stakeholder dialogues and citizens' juries have been grafted on to the ailing body of British science, in the hope that they will give it a new lease of life. Every so often, a few drops of PUS still dribble out from a Lewis Wolpert or a Lord Taverne,[12] but these voices are now a dwindling force. The science community has embraced dialogue and engagement, if not always with enthusiasm, then at least out of a recognition that BSE, GM and other controversies have made it a non-negotiable clause of their 'licence to operate'.

Yet despite this progress, the link from public engagement back to the choices, priorities and everyday practices of science remains fuzzy and unclear. Processes of engagement tend to be restricted to particular questions, posed at particular stages in the cycle of research, development and exploitation. Possible risks are endlessly debated, while deeper questions about the values, visions and vested interests that motivate scientific endeavour often remain unasked or unanswered. And as the GM case demonstrates, when these larger issues force themselves on to the table, the public may discover that it is too late to alter the developmental trajectories of a technology. Political, economic and organisational commitments may already be in place, narrowing the space for meaningful debate.

But now, a new term has entered the lexicon of public engagement. Scientists and science policy-makers are increasingly recognising the limitations of existing approaches, and there has been a surge of interest in moving engagement 'upstream'. In a paper for the 2003 Progressive Governance conference, Demos and Green Alliance explored the case for upstream engagement in the context of nano-technology:

> *Much nanotechnology is at an equivalent stage in R&D terms to biotechnology in the late 1970s or early 1980s. The forms and eventual applications of the technology are not yet determined. We still have the opportunity to intervene and improve the social sensitivity of innovation processes at the*

design-stage – to avoid the mistakes that were made over GM and other technologies.[13]

The language of upstream engagement features in several recent policy statements. The Royal Society's nanotechnology report acknowledges that 'Most developments in nanotechnologies, as viewed in 2004, are clearly "upstream" in nature'[14] and calls for 'a constructive and proactive debate about the future of nanotechnologies [to] be undertaken now – at a stage when it can inform key decisions about their development and before deeply entrenched or polarised positions appear.'[15]

The government appears to agree. Welcoming the Royal Society's findings, Lord Sainsbury, the Science Minister, said 'We have learnt that it is necessary with major technologies to ensure that the debate takes place "upstream", as new areas emerge in the scientific and technological development process.'[16] And most significantly, the government's new ten-year strategy for science and innovation includes a commitment 'to enable [public] debate to take place "upstream" in the scientific and technological development process, and not "downstream" where technologies are waiting to be exploited but may be held back by public scepticism brought about through poor engagement and dialogue on issues of concern.'[17]

A few lessons learned

What has triggered this sudden enthusiasm for upstream engagement? There are a variety of factors and motivations at work. Most immediately, policy-makers and the science community are desperate to avoid nanotechnology becoming 'the next GM'. The wounds of that battle are still raw, and there is little appetite for a rerun. One of the criticisms levelled at the 2003 'GM Nation?' debate is that it took place too late to influence the direction of GM research, or to alter the institutional and economic commitments of key players. The GM–nano comparison should not be applied in an uncritical way, but there is little doubt that the remit of the Royal

Society inquiry, and its call for more upstream debate, was profoundly influenced by GM.

Second, this desire to learn from what has gone before extends beyond GM across the wider realm of biotechnology and the life sciences. It is widely felt that processes of public debate and engagement around human embryology and genetics, from the pioneering work of the Warnock Committee in the 1980s through to the activities of the Human Fertilisation and Embryology Authority and the Human Genetics Commission today, have 'worked' in a way that similar processes around GM have 'failed'.[18] Interesting assumptions lie behind such framings of 'success' and 'failure', but there are instructive contrasts to be drawn. As Sheila Jasanoff notes: 'If the growth of agricultural biotechnologies was marked by too little deliberation, then human biotechnologies seem burdened by almost a surfeit of public soul-searching.'[19] One reason why public debates over human biotechnology are considered to have played out more successfully is that deliberative processes began early and have kept pace with scientific developments.

This connects to a third set of motivations for policy-makers to embrace upstream engagement. The government has placed great emphasis on science and innovation as central pillars of its economic strategy. The foreword to its new ten-year strategy makes the now familiar argument that 'nations that can thrive in a highly competitive global economy will be those that can compete on high technology and intellectual strength. ... These are the sources of the new prosperity.'[20] An extra £1 billion has been allocated to science over the period of the next spending review – a real-term increase of 5.8 per cent each year until 2008. But a big question remains unanswered: Will all of this extra cash, and the innovation it aims to unleash, improve or worsen relations between science and society? The ten-year strategy addresses this dilemma only indirectly, but its advocacy of upstream debate is clearly influenced by a desire to remove obstacles that might upset the innovation apple-cart. Tony Blair was more explicit about this danger in his May 2002 speech to the Royal Society: 'When I was in Bangalore in January, I met a group of academics who were also in business in the biotech field. They said to me bluntly: "Europe has gone soft on science; we are going to leapfrog you and you will miss out".[21]

So, in debates over science and society, a small but significant shift is underway. The sudden vogue for upstream engagement may prove ephemeral, or may develop into something more promising. Andy Stirling, one of the UK's leading thinkers on public engagement, is optimistic. In a recent paper, he predicts that 'New political arenas look set to open up, as "upstream" processes of knowledge production, technological innovation and institutional commitment begin to acquire their own distinctive discourses on participation.'[22]

Yet it is also important not to overstate the novelty of moves in this direction. Sheila Jasanoff describes how in the early years of biotechnology, upstream efforts to identify risks and explore ethical dilemmas were led by the science community itself. In 1973, the US National Academy of Science established a committee under the chairmanship of Paul Berg to explore the potential risks of recombinant DNA research. As Jasanoff notes, 'Thirty years and several social upheavals later, the Berg committee's composition looks astonishingly narrow: eleven male scientists of stellar credentials, all already active in rDNA experimentation.'[23] Nonetheless, the committee's conclusions were precautionary: it called for a voluntary moratorium on certain types of research until more was known about the risks.

At around the same time, the US Office of Technology Assessment was established to provide Congress with 'early indications of the probable beneficial and adverse impacts of the applications of technology'.[24] And in the Netherlands, theorists such as Arie Rip spent much of the 1980s and 1990s developing methods of 'constructive technology assessment' (CTA) for use by the Dutch government, in an effort to embed social values in the design stages of innovation.[25]

Britain was slower to adopt these new techniques, but in 1994, inspired by Dutch models of CTA and Danish use of consensus conferences, the Science Museum and the Biotechnology and

Biological Sciences Research Council (BBSRC) organised a consensus conference on plant biotechnology. This event, held over three days at Regent's College in London, is often cited as the first British attempt at upstream public engagement. In front of an audience of over 300 people, a panel of ordinary citizens took evidence and cross-examined a range of expert witnesses, before coming to their conclusions.[26]

What makes the current talk of upstream engagement any different from what has gone before? Clearly, any new efforts will be informed by and build on these past experiences and methodologies. But if we review these earlier approaches, they appear lacking in several respects. First, they relied primarily on narrow forms of expert knowledge and analysis. More diverse and plural forms of public knowledge were either marginalised (in the case of the Berg Committee or the US Office of Technology Assessment), or implicitly given lower priority (as in the design of the BBSRC's consensus conference, with its reliance on expert witnesses). Second, the framing of debates and the range of issues up for discussion was restricted primarily to questions of risk in the application of new technologies. More fundamental questions around ownership, control and the social ends to which the technology would be directed were ignored.

Most importantly, these initiatives usually took place in a vacuum – with no explicit link back to the research choices and innovation priorities of scientists or industry, or to the decisions of policy-makers. CTA stands out as an exception, as its results flowed into the work of the Dutch government. Elsewhere, the connections with policy were absent from the start, or quickly broken. In the US, the cautious warnings of the Berg Committee were soon lost in the cloud of optimism and hubris that enveloped biotechnology, and in 1995, the Office of Technology Assessment was closed down. In Britain, there was no clear mechanism through which the conclusions of the BBSRC's consensus conference could influence the political and public debate that followed. One commentator describes it as 'an admirable initiative that took place in a political cul-de-sac'.[27]

The purpose of this pamphlet

So, as we embark on a fresh attempt to head upstream, there are lessons to be learnt and mistakes to be avoided. This pamphlet aims to make a practical contribution to that task. Our argument is primarily directed towards UK science policy, but we hope that it also has resonance elsewhere.[28] Our intended audience comprises those with an interest in questions of science and society, but particularly scientists and academics, their professional institutions, the research councils, science policy-makers, R&D-led businesses and NGOs.

As you would expect from two authors with a keen involvement in environmental debates, we also want to explore how these ideas can contribute to thinking about sustainable development. After almost a decade of disagreement over GM, relations between the scientific establishment and the environmental movement have sunk to a depressingly low ebb. Neither side can have lost sight of the paradox inherent in a movement which relies on science to understand the biophysical systems that underpin life on earth (and on technologies such as renewables to help steer us towards sustainability) locked into such protracted conflict with the individuals and institutions that govern British science. Now that an uneasy – and no doubt temporary – truce is in place over GM, we hope that these proposals might contribute in some way to a more constructive climate of openness and collaboration.

In chapter 2, we argue that debates over science and technology, even when they involve processes of public engagement, have been dominated by questions of risk assessment. This framework is too narrow, and fails to ask or answer the more fundamental questions at stake in any new technology: Who owns it? Who benefits from it? To what ends will it be directed?

In chapter 3, we turn to the question of what constitutes a successful engagement process. We explore the reasons for moving upstream, and identify what this might look like in practice. We also discuss the wider implications of public engagement for democracy.

Chapter 4 applies our argument to processes of research and

development within the private sector. We expose some of the tensions between innovation policy and public engagement, and identify resources from within management theory and debates over corporate social responsibility that could be drawn on in resolving those tensions.

Lastly, chapter 5 suggests some practical ways to embed upstream engagement in science, government and society. We examine its implications for public policy, research councils, academia, business, the media and NGOs, and close by drawing out the implications of our argument for the process of science itself. If we take the case for upstream engagement to its logical conclusion, will it not only change the relationship between science and public decision-making, but also the very foundations of knowledge on which science rests?

In science, as in politics, answering one question inevitably raises others. The process of inquiry and learning is endless. We should know that no matter how well we handle one new development, controversy is not about to disappear. We cannot hope to reach the mythical end-point of consensus, the middle ground of prudent progress behind which everyone can rally. The challenge is to recognise that we rely on this constant questioning and the innovation that drives it. Instead of shrinking from scientific and technological endeavour for fear of the uncertainty that accompanies it, we should work to create the conditions for science and technology to thrive. But the simultaneous challenge is to generate new approaches to the governance of science that can learn from past mistakes, cope more readily with social complexity, and harness the drivers of technological change for the common good.

This pamphlet suggests one such approach. To return to the idea of science as performance, the task of upstream engagement is to remove some of the structures that divide the back-stage from the front-stage. It seeks to make visible the invisible, to expose to public scrutiny the values, visions and assumptions that usually lie hidden. In the theatre of science and technology, the time has come to dismantle the proscenium arch and begin performing in the round.

2. Science and the social imagination

Reports that say that something hasn't happened are always interesting to me, because as we know, there are known knowns; there are things we know we know. We also know there are known unknowns; that is to say we know there are some things we do not know. But there are also unknown unknowns – the ones we don't know we don't know.

Donald Rumsfeld[29]

Chris Patten came a close runner-up with his observation that 'having committed political suicide, the Conservative party is now living to regret it'. But the 2003 award for most absurd remark by a public figure went to that guru of obfuscation, Donald Rumsfeld. The US Secretary of Defense won the Plain English Campaign's 'Foot In Mouth' trophy for his 62-word attempt to clarify a point at a NATO press briefing. His intention was to defend Washington's view that the US could not wait for 'absolute proof' before taking action against groups and states suspected of acquiring weapons of mass destruction, but his remark left NATO allies and journalists completely baffled. 'We think we know what he means', said John Lister, a spokesman for the Plain English campaign, 'but we don't know if we really know.'[30]

Yet for once, Donald Rumsfeld is on to something. Political processes, when confronted with advances in science and technology, are generally incapable of dealing with anything beyond *known* uncertainties. They can only address questions that they already know how to ask. Conversations are framed in a way that denies or edits out unpredictable consequences. This is despite a growing body of evidence that it is these same areas of ignorance and ambiguity that are of greatest public concern.[31]

Researchers at Lancaster University compared public attitudes towards GM and information technology. They found that controversies arise in areas where, although it is sensed that there is no knowledge, this is conveyed misleadingly in terms of reductionist scientific uncertainty. In such situations, the research team identified a 'deep cultural dislocation' between the way that policy-makers and the public frame relevant questions. 'Whilst the former tend to ask simply "What are the risks?", the latter ask in addition, "What might be the unanticipated effects? Who will be in charge of, and will take responsibility for, the responses to such surprises? And can we trust them?"'[32] Many public engagement processes, however well-intentioned, get caught in this trap. Questions of risk – the known uncertainties – can easily dominate proceedings and squeeze out broader discussion of unknown or unanticipated consequences.

The tyranny of risk assessment

Why is it that whenever a new development in science or technology sparks debate, the key elements of that debate are then framed by scientists and policy-makers as 'risk issues'? Michael Power argues that there is now an overwhelming tendency in political and organisational life to reach for 'the risk management of everything'.[33] In a recent Demos pamphlet, he describes how risk management, once an obscure and technical practice within the private sector, has become a dominant discourse within public service delivery and at 'the heart of government itself'.[34] This is reflected in a 2002 report by the Prime Minister's Strategy Unit, which subsequently gave rise to a set of risk management principles for all government departments.[35]

If we accept Power's analysis, then the centrality of risk questions to science and technology debates is part of a much wider trend. But in the 'risk society',[36] perhaps the biggest risk is that we never get around to talking about anything else. Even the 'new mood for dialogue' identified by the House of Lords in 2000 has struggled to alter this dynamic. Brian Wynne describes how the past five years have seen a huge flowering of practical and analytical work aimed at nurturing dialogue and mutual understanding between science and society. Yet the 'radical apparent potential' of these activities 'is compromised by deeper, less manifest cultural assumptions and commitments . . . [which] have yet to be identified, confronted and changed.'[37]

Wynne pinpoints two factors that contribute to this problem. The first is that most forms of public participation are focused on downstream risks or impacts, 'reflecting the false assumption that public concerns are only about instrumental consequences, and not also crucially about what human purposes are driving science and innovation in the first place'. The second is an assumption that the task of defining what the salient issues are within processes of public engagement automatically falls to experts, leaving citizens with 'no capability nor proper role in autonomously creating and negotiating . . . more diverse public meanings.'[38]

In the next chapter, we explore how engagement processes might overcome these limitations. Our argument here is not that involving the public in risk assessment is a bad idea. Quite the opposite: any process of evaluating risk and designing responses to it is likely to be greatly enriched by public involvement. There is a rich literature on participatory risk assessment, and organisations such as the Environment Agency – our partners in this pamphlet – have pioneered the use of these methodologies in their work.[39]

Yet when we are faced with potentially disruptive innovations, the danger is that risk assessment – however participatory – merely digs us deeper into the hole that we are trying to escape from. It avoids our real predicament, which is one of ignorance and ambiguity. Debates are too often framed in terms of 'Is it safe?', with the implication that the likelihood of certain outcomes is susceptible to rational calculation. More challenging questions which flow from ignorance about the long-term social consequences of a technology's development are never asked, let alone answered.

This concentration on risk is an entirely understandable way of rationalising an otherwise open and daunting set of questions. It reflects what Zygmunt Bauman memorably describes as modernity's 'gardening instinct'.[40] Yet this desire to tidy the borders and prune the hedges of our democracy means that many public engagement processes are stripped of any meaningful content. Sheila Jasanoff recalls how this process played out in the development of GM technology:

> *Within barely a decade, environmental consequences that were once considered speculative and impossible to assess came to be regarded within policy circles as amenable to rational, scientific evaluation. By 1990, it appeared that, for genetically modified crops, apocalyptic visions and the rhetoric of science fiction could be set aside in favour of objective expert discourses and routine bureaucratic approvals.*[41]

Brian Wynne goes so far as to argue that 'virtually all of the mushrooming commitment to public citizen engagement in 'science policy' . . . is something of a mirage'.[42] One deficit model has gone, only to be replaced by another – a misunderstanding of what is at stake and what is the basis of public concern. Recently, in the context of international development, the relentless drive for participation has been dubbed 'the new tyranny'.[43] Here, the process of taking questions that are essentially political and reducing them to issues of risk management has a tyrannical aspect of its own.[44]

Downstream, the flow of innovation has absorbed numerous engagement processes. Yet few of these have any real connection to the upstream questions that motivate public concern: *Why this technology? Why not another? Who needs it? Who is controlling it? Who benefits from it? Can they be trusted? What will it mean for me and my family? Will it improve the environment? What will it mean for people in the developing world?* The challenge – and opportunity – for upstream public engagement is to force some of these questions back on to the negotiating table, and to do so at a point when they are still able to influence the trajectories of scientific and technological development.

This process will also require the tacit visions and assumptions that shape research priorities to be surfaced. The philosopher Mary Midgley has written at length about the 'myths, metaphors, images and the other half-conscious apparatus' that form part of the workings of the scientific imagination.[45] Yet despite their importance, such visions – or 'imaginaries' in the jargon of social science – are also squeezed off the agenda by risk-based approaches to dialogue, and are rarely opened to wider public scrutiny.

Who needs it?

Ten years ago, the consultancy SustainAbility published a report called 'Who Needs It?', which sought to introduce a new set of questions into the corporate sustainability arena.[46] It argued that simply adding a dash of eco-efficiency or a drop of social responsibility to the existing mix of products and services was no longer sufficient. The markets of the future would be shaped by human values and needs that until now have been unarticulated and unserved. Key characteristics of this impending values shift would include a new focus on inter-generational equity, and a desire to meet the basic needs of people in the developing world. This would be good news for certain sectors, whose offerings would adjust well to the new social climate, but bad news for others, who would eventually discover that they no longer chimed with the expectations of consumers. To help determine the winners and losers in this brave new world, the report proposed a 'needs test' that companies should carry out every time they proposed a new product or service.

Surveying the business landscape in 2004, the values shift predicted by SustainAbility still looks to be some distance away. But their report alerts us to some important questions that are highly relevant to our analysis. With new technologies, the question 'Who needs it?' is very rarely asked. Means have an awkward habit of becoming ends. There is often a circularity in the arguments used to support new developments – just because a new technology is *possible,* it is therefore seen as desirable. Before a proper conversation can get underway about the priorities and ends to which the technology should be directed, policy-makers have already skipped on to the next layer of questions about how to deal with the risks, benefits and consequences of its exploitation. Policy and regulatory debates tend to assume that the debate about ends has already occurred – that the economic and social benefits of innovation are obvious and agreed. But of course, this is rarely the case.

Similarly, in all the excitement that surrounds new technologies, it is easy to neglect the untapped potential of the technologies we already have at our disposal. Our list of unasked questions grows longer: 'What are the outcomes that this technology seeks to generate? Could we get there in another, more sustainable and cost-effective way?' We should never underestimate the power that the techno fix exerts over the political imagination. But as governments everywhere are learning through bitter experience, environmental problems will not be solved by technological innovation alone. It is easy to throw money at the technological end of the problem. But this must be accompanied by social and political innovation that alters the frameworks within which choices are made. Charles Leadbeater makes this point well in the context of transport:

> *New, more sustainable forms of car transport will require scientific and technological innovation, such as new fuel sources for cars. But the true potential will not be realised without social innovation to create new patterns of car use, and even ways for consumers to share and own cars through leasing schemes. It will require regulatory innovations such as road pricing, which may well only be possible if we have political innovations to give cities more powers to control their own transport taxation. We need to imagine not just new technologies, but whole new social systems for transport.*[47]

A similar point can be applied to energy systems. Our current model of energy production is not the optimal arrangement – it is simply the product of a series of decisions made over a long period. Vijay Vaitheeswaran, in his investigation into the history and prospects for renewable energy, identifies a precise moment back in 1884 when we unwittingly locked ourselves into the modern, centralised, grid-based electricity system. He explains that early experiments in electrification in the US, led by Thomas Edison, centred on stand-alone 'micropower' plants in homes and offices around New York City. A fight for investment between Edison and his rival Nikola Tesla, which Edison eventually lost, meant that grid electrification based on AC technology became the backbone of the electricity network.[48]

As this story illustrates, the energy system we have is not the only way of doing things. A return to Edison's micropower vision – updated, of course, with the renewable technologies we now have at our disposal – offers one route to solving the energy crisis and climate change. Yet a micropower system is very different, socially and institutionally, as well as technologically, from our current system. We cannot simply rejig a few financial incentives and expect micro-renewables to win out. We need to look not just at technology policy, but also at issues of planning, land use and building design. In other words, we need a form of 'whole-system innovation'.[49]

Similar stories could be told about waste, water policy or agriculture. In each of these areas, technology is part of the solution, but it is no panacea. No one has made this argument better than Fritz Schumacher, in his classic text *Small is Beautiful*. Schumacher describes how the 'forward stampede' advocating new technologies 'burst into the newspaper headlines every day with the message, "a breakthrough a day keeps the crisis at bay".[50] This constant focus on 'breakthroughs' distracts attention from the real, though not very technological, problems that we face – not to mention the real, though not very technological, solutions which lie within our grasp. Schumacher recognised this only too well. As he wrote, 'it takes a certain flair of real insight to make things simple again'.[51]

The politics of small things

As we have seen, two impulses within existing models of public engagement threaten to undermine their radical potential. The first is for policy-makers and other experts to restrict the space for debate to a technocratic discourse around risk. The second is for deeper questions about human needs and ends to be squeezed off the agenda. Both of these tendencies were evident during the GM controversy. But will the same apply to emerging debates over nanotechnology?

The Royal Society's nanotechnology report is a good place to start in considering this question. Here we find several encouraging signs of a new approach. A whole chapter is devoted to social and ethical issues, and another to public dialogue. The latter even includes the language of upstream debate. Yet viewed in its entirety, the body language of the report still signals that questions of risk take priority. Much of its analysis is devoted to these, along with a large share of the recommendations. And although the report acknowledges that social concerns are likely to focus on two questions – 'Who controls use of nanotechnologies?' and 'Who benefits?' – little attempt is made to follow through and answer these, beyond a call for more research.[52] There is no real sense that these questions are up for serious negotiation within the terms of reference of the inquiry.

This is perhaps understandable. As a recent paper from Lancaster University and Demos argues, when faced with new situations policy-makers generally turn to the tools and frames of reference that lie close at hand. Just as early policy discussions around GM were shaped by risk assessment models that were originally developed within the nuclear industry, so discussions around nanotechnologies are likely to inherit models that were devised for GM. The way such patterns repeat themselves highlights the need for a more searching analysis of the distinctive character and properties of nano-technologies before regulatory commitments are made. 'It cannot be assumed that the conceptualisations and analytical categories currently available will be able to capture what may prove to be most distinctive about nanotechnology. In other words, be very careful to ensure we don't set ourselves up to fight the last war.'[53]

In the same way, the visions and 'imaginaries' of nanoscientists need to be brought to the surface and opened up to wider debate. One doesn't have to delve far into the scientific and policy literature around nanotechnology to find enormous claims being made about its transformative potential. Bottom-up or top-down, the promises of nanotechnology are revolutionary. For its growing band of cheerleaders in government, academia and industry, nanotechnology offers unlimited energy, targeted pharmaceuticals and intelligent materials. It is increasingly talked about with the same breathless enthusiasm that surrounded biotechnology and information technology in the mid-1990s. Take for example the opening paragraph of a recent parliamentary report:

Nanotechnology is more than an exciting new technology. . . . Over the coming years and decades, nanotechnologies are set to make an enormous impact on manufacturing and service industries; on electronics, information technology; and on many other areas of life, from medicine to energy conservation. . . . Nanotechnology has been described as a new industrial revolution.[54]

Nanovisions

For most people, nanotechnology is still an unknown quantity. But swirling around behind the science are many different views of its social implications and transformative potential. These fall into at least three categories:[55]

o **Nano-radicals** see nanotechnology as profoundly disruptive of economies and societies. In his 1986 book *Engines of Creation*, Eric Drexler, the so-called 'father of nanotechnology', predicted a world in which nanoscale machines – 'molecular assemblers' – would be capable of arranging atoms to build almost anything from the bottom up. Because it would take millions of these assemblers to build anything, Drexler argued that assemblers would also need to be capable of replicating themselves, hence his famous – and now disowned – scenario of self-replicating nanobots smothering the world in 'grey goo'.[56]

o **Nano-realists** emphasise the incremental innovations and commercial returns that the technology will provide in sectors such as manufacturing, IT and healthcare. They aren't interested in the hypothetical possibilities of bottom-up molecular manufacture. Theirs is a venture-capitalised, research-council approved version of nanotechnology, focused on practical applications and economic returns. It is this vision that has excited policy-makers and unleashed a cascade of government funding across the industrialised world.

o **Nano-sceptics** count Prince Charles and Michael Crichton among their number, but their most active and articulate representatives are the ETC Group, a small Canadian NGO. It's not 'grey goo' that worries them so much as the immediate risks posed by nanoparticles to human health and the environment. They also have some pretty serious questions about who is controlling the technology and whose interests it will ultimately serve.

Such categorisations inevitably simplify a complex range of perspectives, but they are useful in understanding how public perceptions of nanotechnology could evolve over time. Crucially, these underlying visions also inform and shape the direction of scientific research.

In the United States, a report from the National Science Foundation paints a vivid picture of the changes that will result from the convergence of nanotechnology with biotechnology, information technology and cognitive science. This deserves to be quoted at some length:

Developments in systems approaches, mathematics and computation . . . allow us for the first time to understand the natural world, human society, and scientific research as closely coupled, complex, hierarchical systems. At this moment in the evolution of technological achievement, improvement of human performance through integration of technologies becomes possible. Examples of payoffs may include . . . revolutionary changes in healthcare, improving both individual and group creativity . . . brain-to-brain interaction, perfecting human–machine interfaces . . . and ameliorating the physical and cognitive decline that is common to the aging mind. . . . Moving forward simultaneously along many of these paths could achieve an age of innovation and prosperity that would be a turning point in the evolution of human society.[57]

Contained within such projections is a set of assumptions about future human and social needs that are contestable and should be debated. The wider public needs to turn its back for a second only, and the slide from means to ends is underway. The mere *possibility* that nanotechnology may enable certain changes to the lives of individuals, families and communities is seen as sufficient justification for those changes to go ahead. It is important to understand how these underlying visions inform and shape the direction of scientific research. Tacit visions or 'imaginaries' of the social role of nanotech form the basis on which research priorities are negotiated and planned. This 'sociology of expectations' is now attracting wider interest with the field of science and technology studies.[58]

In 2005, the National Science Foundation, which published this study of 'converging technologies and human performance', holds the purse strings of a budget of $5.7 billion. The democratic deficit in such processes of framing, prioritisation and resource allocation thus takes on a more alarming aspect. Who decided that these were the ends to which technological convergence should be directed? On what authority, and with what processes of public consultation? Why were 'brain-to-brain interaction' and 'human–machine interfaces' selected as priorities for research? To what extent do these developments represent 'techno-fix' solutions to problems that might be addressed better in other ways? And in a world where more than one billion people lack access to safe water and millions die each year from preventable disease, are these really the best uses we can make of billions of research dollars?

The GM saga shows what can happen when the underlying social visions of key players (such as Monsanto) are not made visible and opened up to public deliberation. The challenge now facing those involved in nanotechnology research is to approach things differently: to articulate the visions, promises and expectations of the technology at an earlier stage, and make them the focal point of upstream public engagement.

3. The rules of engagement

The Great Yorkshire Showground in Harrogate, famed for its agricultural shows, is an unlikely setting for an exercise in democracy. But one afternoon last summer, 250 people gathered there to voice their hopes and fears about genetically modified crops and foods. Bruised and weary from its conflicts with the press, public, pressure groups and scientists, the government decided to confront the issue head-on and sponsor the *GM Nation?* debate. This was an innovative attempt at public engagement, and may come to be seen as marking a sea change in the government's approach to science and technology. In agreeing to a public debate on GM, ministers were implicitly acknowledging the inadequacies of previous attempts to handle such issues, and signalling their intention to try a new approach.

It can be hard to get these things right the first time – and there was certainly plenty of criticism of the *GM Nation?* process. An independent evaluation uncovered a number of shortcomings, including inadequate resourcing, a failure to engage members of the public who had not previously been involved in GM issues, and a lack of space for genuine deliberation.[59] But perhaps the biggest flaw of the *GM Nation?* process was its timing – it took place too late to influence the direction of GM research, or to alter the institutional commitments of the biotechnology industry and other key players.

But despite these failings, *GM Nation?* points the way to a more sophisticated and deliberative handling of such issues by government.

This chapter reflects on the lessons of *GM Nation?* to ask what an ideal upstream engagement process should look like. We want to seize the gauntlet laid down by the Royal Society's nanotechnology report: 'to generate a constructive and proactive debate about the future of the technology now, before deeply entrenched or polarised positions appear.'[60] We start with the fundamental question of *why* engagement is important and what purposes it can serve. We move on to look briefly at *how* to engage: what methods can be used to involve people in decisions? And finally we ask *is engagement enough*? Can processes like *GM Nation?* help us to shape new technologies better, or do we need to make more fundamental changes to the way that governments work? In other words, how do deliberative processes fit within wider trajectories of political or regulatory decision-making?

Why engage the public?

When commissioning *GM Nation?*, the government was strangely silent on perhaps the most important question – why it decided to run the process, and what it intended to do with its findings. The official reason for the debate was that the government's advisers, the Agriculture and Environment Biotechnology Commission, had asked for it. But it was never made clear how the findings would be used in future decisions on GM. Would the government accept the verdict of the debate and follow its recommendations? Would it support or oppose the commercialisation of GM crops on the basis of the evidence received? These issues were not at all clear. As the National Consumer Council observed at the time, 'The impression created was of consultation without inclusion, raising questions about whether the government genuinely had an open mind.'[61]

A clear lesson from *GM Nation?* is that the objectives of any public engagement process should be clear from the start. It might simply be designed to gather information about public opinion – or, at the other end of the spectrum, to determine a policy decision. In disentangling the different reasons for public engagement, a useful distinction can be made between *normative, instrumental* and *substantive* motivations.[62]

The *normative* view states that such processes should take place because they are the right thing to do: dialogue is an important ingredient of a healthy democracy. The *instrumental* view holds that engagement processes are carried out because they serve particular interests. Companies developing a new technology may want to find out what people think, so that they can present their innovation in the best possible light. Governments may want to engage in order to build trust in science and manage their reputation for competence.

From a *substantive* perspective, engagement processes aim to improve the quality of decision-making, to create more socially-robust scientific and technological solutions. The goal is to improve social outcomes in a deeper sense than just improving the reputation of the technology, company or government involved. From this point of view, citizens are seen as subjects, not objects, of the process. They work actively to shape decisions, rather than having their views canvassed by other actors to inform the decisions that are then taken.

With hindsight, it seems that the motivation behind *GM Nation?* was partly normative – the government wanted to do the right thing, as recommended by its advisers. It was definitely instrumental – ministers wanted to be *seen* to be doing the right thing, in order to build trust in their handling of the issue, and perhaps to move towards greater acceptance of the technology. But given that it was never made clear how the results of the debate would be used, it seems unlikely that there was any substantive motivation behind the debate. It was not aimed at making better, more informed decisions about GM. In this respect, it was too little, too late.

As Andy Stirling points out, substantive approaches are particularly important when there are 'intractable scientific and technological uncertainties . . . as a means to consider broader issues, questions, conditions causes or possibilities.'[63] Although there is still a role for normative and instrumental approaches, it is clear from the GM debate, and from emerging debates about nanotechnology, that public engagement must be substantive. It must not just inform decisions – it must shape them.

Opening up rather than closing down

Stirling goes on to make a helpful distinction between processes that aim to *open up* a debate, and ones that aim to *close it down*. For engagement to be meaningful, it needs to open up some of the deeper questions discussed in chapter 2 – to look at who frames the visions and purposes of a new technology, and to allow the public to ask the questions that they consider most important.

Stirling reminds us that implicit assumptions lie behind all engagement processes. Supposedly 'objective' reports by 'experts' are actually carefully and subjectively framed according to the outlook of the experts themselves, and the questions they choose to answer. This is where the retreat to a risk discourse occurs – with biotechnologists, for example, choosing to answer the question 'is it safe?', rather than 'is it necessary or desirable?'. Identical framing assumptions apply to engagement processes. Decisions made about the type of process used, the participants, the questions asked, the information provided, and so on can lead to inadvertent bias or deliberate influence.[64] In other words, you can get the results you want if, consciously or subconsciously, you frame the debate in the right way.

Even engagement processes such as citizens' juries or consensus conferences can be used to close things down, in just the same way as risk assessment. The worst outcome would be one in which techniques for engagement are incorporated into the bureaucratic processes of decision-making without changing the way that decisions are made. In this case, 'public engagement' is no more than a process box that civil servants and scientists have to tick when drawing up a policy or applying for funding.

By contrast, practised in a meaningful way, public engagement can lead to better, more robust policy and funding decisions, provided it is used to open up questions, provoke debate, expose differences and interrogate assumptions. From this perspective, it is not up to 'experts' to frame a question and slot in an engagement process to provide the answer. As Brian Wynne has argued, this is simply the deficit model in a new guise. Instead, the public should help to decide the questions and the way in which a particular issue will be approached.

There is a hint of this 'mark-2' deficit model in the Royal Society's nanotechnology report. The need for engagement processes is flagged alongside other tasks – additional research, risk assessment, life-cycle analysis, and so on. The Royal Society has written a recipe for the public consumption of nanotechnologies, in which one of the ingredients is public involvement – alongside numerous others. A more substantive model of engagement would hand over authorship of the recipe to a more plural and diverse set of publics, rather than reducing the public to just another ingredient in the pot. This more substantive model of engagement would genuinely 'open things up' in Stirling's terms, by uncovering framing assumptions and making citizens the subjects rather than the objects of the process. Engagement would then be overarching, rather than bolt-on.

How to engage?

Once the question 'why engage?' has been answered, we can then turn to the secondary question of 'how to engage?' – what methodologies allow a proper consideration of public views and values? A great deal of energy has been expended on this in recent years, but the short answer is that there is no ideal process, but a menu of different methods and techniques. From focus groups to referendums, citizens' juries to stakeholder dialogue, there are as many processes for engagement as there are issues to debate. Our argument here is that aim should come before method, but it is worth reviewing the different techniques on offer. The box below shows some – though by no means all – of the methods available.

Methods of public involvement[65]

Deliberative polling
In a deliberative poll, a large, demographically representative group of perhaps several hundred people conducts a debate,

usually including the opportunity to cross-examine key players. The group is polled on the issue before and after the debate.

Focus groups

A focus group is a qualitative method used widely in commercial market research and increasingly in academic social research. Typically, a group of eight to ten people, broadly representative of the population being studied, is invited to discuss the issue under review, usually guided by a trained facilitator working to a designed protocol. The group is not required to reach any conclusions, but the contents of the discussion are studied for what they may reveal about shared understandings, attitudes and values. Focus groups may also help to identify the factors (which large-scale surveys rarely do) that shape attitudes and responses, including trust or mistrust. They also help in the design and interpretation of quantitative public opinion surveys.

Citizens' juries

A citizens' jury (or panel) involves a small group of lay participants (usually 12–20) receiving, questioning and evaluating presentations by experts on a particular issue, often over three to four days. At the end, the group is invited to make recommendations. In the UK to date, local authorities, government agencies, policy researchers and consultants have convened over 200 citizens' juries on a wide range of policy issues.

Consensus conferences

By convention, a group of 16 lay volunteers is selected for a consensus conference according to socioeconomic and demographic characteristics. The members meet first in private, to decide the key questions they wish to raise. There is then a public phase, lasting perhaps three days, during which the group hears and interrogates expert witnesses, and draws up a report. The main differences between a consensus conference and a citizens' jury or focus group are the greater opportunity for the participants to become more familiar with the technicalities of the subject, the greater initiative allowed to the panel, the admission of the press and the public, and the higher cost.

Stakeholder dialogues
This is a generic term applied to processes that bring together affected and interested parties (stakeholders) to deliberate and negotiate on a particular issue. Stakeholders can range from individuals and local residents to employees and representatives of interest groups.

Internet dialogues
This term is applied to any form of interactive discussion that takes place through the internet. It may be restricted to selected participants, or open to anyone with internet access. The advantages of internet dialogue include the ability to collect many responses quickly and to analyse them using search engines. Similarly, they can combine the benefits of rapid exchange of ideas (brainstorming) with a complete record. On the other hand, participation may be self-selecting and unrepresentative, and the anonymity of the internet may encourage impulsive rather than considered responses. Anonymity may make it difficult to investigate the provenance of information provided.

Deliberative mapping
This is a process in which expert and citizen assessments are integrated. In a deliberative mapping exercise, citizens' panels and specialist panels are convened and interact with each other, allowing participants to interrogate each others' views and knowledge, and exposing framing assumptions made by both sides. Deliberative mapping seeks to bring together the views of 'experts' and 'public', through face-to-face deliberation between these two groups. The approach was pioneered through a consortium of research institutes in the UK, and applied to the specific problem of organ transplant options.[66]

The type of process used will, obviously, depend on what is required. Timing and resource constraints will determine how ambitious or far-reaching a process can be. But there are other issues to take into account. Our analysis of normative, instrumental and substantive motivations gives rise to a number of further questions:

○ *Deliberative or snapshot?* Is the process designed to involve people in a process of deliberation, whereby information is processed, and views formed and discussed? Consensus conferences and citizens' juries allow this. Or is the aim merely to get a snapshot of people's views, in order to inform decisions? In this case, a straightforward opinion poll or focus group may be more appropriate.

○ *Representative?* Different methods will be required if the aim is to involve a representative sample or a particular segment of the population. But even smaller, more deliberative processes can be 'representative' in a less formal, statistical way, if participants are selected according to certain criteria – as is routinely the case with focus groups, for example.

○ *Hierarchical or non-hierarchical?* One important, though often overlooked, factor is how 'expert' knowledge is treated. In other words, does the method follow the traditional hierarchy whereby experts decide what questions need addressing, and what information should be taken into account? Or does it subvert this, allowing lay participants to frame questions, gather information and question evidence? The citizens' jury is perhaps the best example of a process in which hierarchies are reversed, with jury members free to define the question, call witnesses and seek whatever information they deem relevant or necessary. Such a model makes it easier to open up rather than close down debates, along the lines discussed above.

○ *Consensual or exploratory?* Lastly, it is important to

consider whether a deliberative process aims to reach a consensus, or simply to explore views. Some processes, such as deliberative polling, aim to develop a richer understanding of views. Consensus conferences and stakeholder dialogues often aim to bring about consensus, or even to reach a definitive decision.

Are engagement processes enough?

So far, we have explored questions of why and how to engage. But true to the spirit of opening up, our answers raise larger questions about the nature of democracy. Where does public engagement fit in the set of relationships between citizen and government?

This may seem far removed from the immediate and practical question of how to handle nanotechnology or GM foods. But as we saw in the last chapter, the question of how to handle new technologies is not a technical or procedural question – it is a question of politics. Decisions about the relationship between technology and society are deeply political. They require forms of mediation between different interests, values and world views. The challenge is how to integrate engagement processes into wider patterns of political decision-making. As Sheila Jasanoff says,

> *The purpose is to hold science and industry answerable, with the utmost seriousness, to the fundamental questions of democratic politics – questions that have fallen into disuse through modernity's long commitment to treating science as a realm apart in its ability to cater for society's needs: Who is making the choices that govern lives? On whose behalf? According to whose definitions of the good? With what rights of representation? And in which forums?*[67]

Simply slotting deliberative processes into existing ways of doing things will not result in any real change. Some of the more naïve proponents of public engagement seem to assume that the way to resolve difficult issues is by bringing together the concerned parties, adding a mix of methods and a family pack of post-it notes, and then allowing the facilitators to save the day. But decisions about the way that technology and society interact are deeply political, and engagement processes need some kind of link to the political system. At the end of the day, decisions have to be made, and elected politicians usually have to make them.

Towards a deliberative democracy

This takes us to the heart of political theory. Advocates of deliberative democracy provide important insights into the relationship between engagement processes and democratic decision-making.[68] They stress, above all, the process by which views are formed. It is often assumed, particularly by politicians and economists, that an individual's political decisions or economic choices are a manifestation of innate beliefs or preferences. If this is true, then a simple opinion poll – or indeed, a referendum – to assess people's views is the easiest way to understand public attitudes. Everyone, the argument goes, will have a view on a particular technology – GM for example – we just need to find out what it is.

By contrast, a deliberative model emphasises that people's views are shaped by the way they encounter or engage with an issue. So people do not have a view on GM unless they are required to have one – which could happen when they read about it in the papers, are asked to buy GM food in a supermarket, are asked by a pressure group to oppose it, or are invited to participate in a *GM Nation?* meeting in their town hall.

This helps us to understand people's reactions to technologies. It is hardly surprising that so many respondents to *GM Nation?* expressed 'unease at the perceived power of the multinational companies which promote GM technology'.[69] This is not a criticism of the technology itself, but of the way it was handled. We begin to see the emergence of a two-way, shifting dialogue: the formation of technologies on the one hand, and the formation of views on the other. The mingling of these two processes is what will ultimately determine the direction of a technology – and society's reaction to it. An important lesson has emerged from this analysis. If it is possible to shape the process of view formation through deliberation, then it is possible for people's views, and the technology they are reflecting on, to be shaped simultaneously. It is no longer a case of developing a technology and finding out what people think about it – the two can, and should, be done in parallel.

In the final analysis, the buck for decisions over science and technology must stop with elected politicians. This is one of the things that we elect them for. But political decision-making should not take place in a vacuum. Rather, it should seek out and take account of diverse forms of social knowledge and intelligence, and use deliberative processes to better inform its decisions.

A final argument in favour of such an approach is that it could help reinvigorate a wider enthusiasm for politics. Critics of engagement processes often point out that political *dis*enagagement is at an all time high. People are not exactly queuing up to be involved in debates about technology. Many choose not even to vote at a general election.[70] Ironically, though, this rejection of politics could be resulting from too little, rather than too much, engagement. Too many people see politics as separate from their everyday lives. If faith in the institutions of representative democracy is on the wane, now is the right time to start experimenting with new forms of democratic debate.

4. Open innovation

A few months ago, the RSA's Forum for Technology, Citizens and the Market published some research that makes sober reading for supporters of public engagement in science. Based on interviews with managers in 12 innovation-based companies, it found low levels of awareness of the need for public engagement and even lower levels of action. When attempts are made to have a dialogue with the public about new innovations they tend to occur long after the key business decisions have been taken. The deficit model is alive and well, with several companies equating engagement with PR-led communication. As one manager put it, '. . .We tend not to publicise too much about what we do until we're actually 101% sure. You could perhaps accuse us of being slightly risk averse.'[71]

Such findings represent a serious challenge to our argument. Moving public engagement upstream is hard enough in the context of taxpayer-funded – and publicly-accountable – science. How can it possibly work in the private sector? Several obstacles stand in the way: the profit motive; pressures for commercial confidentiality; and tight frameworks of patent and intellectual property law.

But there are two reasons why more public engagement in corporate science is needed. First, this is where the lion's share of R&D takes place. In the UK, the private sector research budget is almost double that of the public sector, and the Treasury is relying on business to match it every step of the way as it ratchets up science spending over the next decade.[72] Second, corporate science, and the perceived conflicts of interest at play within it, are the focus of genuine public concern. This is one of the most striking conclusions to emerge from research into public attitudes towards GM.[73]

Yet this message is not being heard. The sector where public engagement is most urgently required is barely engaged with this agenda. The 'new mood for dialogue' around science and technology that the House of Lords identified in 2000 does not appear to have spread to Britain's boardrooms or corporate R&D facilities.

There are honourable exceptions. For example, BT has consistently promoted public discussion of the social, ethical and environmental dimensions of digital technologies.[74] And in the formative stages of the GM debate, Unilever participated in dialogue processes and funded much-needed social research.[75] But a surprising number of science and innovation-led companies remain an engagement-free zone. The emerging debate around nanotechnologies is a case in point: so far, no UK company has shown leadership on this issue, or attempted to contribute to a wider public debate.

Are there any insights into this problem that we can draw from government innovation policy? In the UK's new ten-year strategy for science, there is little effort to link innovation and public engagement in anything other than defensive terms. Public disquiet must be taken seriously, but only in so far as it threatens the smooth passage of new technologies from the laboratory to the marketplace. The language of 'upstream engagement' is there, but to employ the distinction of the previous chapter, the motivations for doing it are instrumental rather than substantive.

Within the ten-year strategy, as Tom Macmillan has argued, the potential links between theories of innovation and arguments for engagement are not fully realised. Engagement is seen as 'an add-on to processes of knowledge creation, rather than an integral part of them', and there is little recognition that non-scientific forms of public knowledge can add new forms of economic and social value to science.[76] As a result, the chapter in the strategy on 'science and society' feels disconnected from many of the initiatives outlined in the rest of the document.

To give one example, there are several references to the development of a new 'Technology Strategy' which will be steered by a board 'comprising mainly senior business leaders' and 'expertly informed through engagement with stakeholders in the science base and business to provide clear and transparent guidance to Government in setting funding priorities.'[77] If the new commitment to upstream public engagement is at all meaningful, then surely the creation of this board is one of the first places to start? Yet the implication is that this new strategic body – with the responsibility for allocating up to £178 million of public money – will be narrowly constituted from business and other expert interests, with little or no space for meaningful public engagement.

Learning from Finland

If policy cannot provide a clear rationale for upstream public engagement, what else can progressive companies draw on to build a business case? One option is to draw on recent developments in management theory that explore a shift to more open models of innovation.[78] Such work suggests that companies should combine external and internal forms of knowledge into new architectures and systems. Charles Leadbeater suggests that open innovation flows from three kinds of social interaction:

1. being good at seeking out or attracting diverse ideas, which clash, collide and spark with one another;
2. being good at absorbing these different insights and combining them with your own knowledge and expertise, to create a new product, service or technology; and
3. excelling at innovating in use, with active consumers who increasingly want a say in how products and services are used.[79]

It is not only companies that can adopt these new patterns of open innovation. Countries can do it too. Leadbeater gives the example of Finland, which despite its size has become one of the most technologically advanced societies in the world. Most of its GDP is derived from electronics, computing and telecoms, and it has given rise to two of the biggest challengers to US corporate dominance: Nokia and Linux. The best way to understand what has made Finland so successful is to see it as an open innovator. Because it is a small country with few resources, it has always had to borrow ideas from abroad. It also has a strong culture of citizen-led innovation: the open source 'hacker ethic' that underpins the growth of Linux.

It doesn't require much imagination to see how such ideas could be applied to public engagement in science. Companies that are serious about open innovation will have a strong motivation to include public participation in their upstream R&D, in order to benefit from the different values, perspectives and forms of social intelligence that this could bring. The wider uptake of open innovation models offers one route to re-energising business efforts at public engagement.

Corporate social innovation

Another area with promise is the growing body of theory and practice around corporate social responsibility (CSR) and sustainability. As with government, the case for public engagement can be made here in different ways. From an instrumental perspective, conjuring up the example of Monsanto is usually enough to make the point that the failure to participate openly in, or the intention to subvert, processes of public debate around new technologies can have disastrous consequences for the profitability and even survival of a firm.

From a substantive perspective, the more interesting question is how public engagement can help companies to create new forms of social value. As Demos has argued elsewhere, the next phase of the CSR debate will need to be based around stronger alliances with the innovation agenda.[80] While CSR has made great inroads in many businesses, it is still held back by two factors:

o   *Marginalisation* – Far too often, CSR and sustainability activities are bolted on to the communications and public affairs departments, and remain removed from the strategic and R&D functions of the business.

O *Bureaucratisation* – The audit-based, box-ticking culture of reports, league tables, and standards are such a dominant focus of many companies' CSR and sustainability efforts that they are in danger of stifling wider forms of social innovation and creativity.

We need to find ways of connecting sustainability and CSR to the core innovative capabilities of an organisation – the R&D, product development and strategy functions that represent the greatest sources of business value. If the knowledge and capabilities of these parts of any business could be applied to social – as well as economic – dimensions of innovation, then the CSR agenda might start to have a real impact. In addition, marketing departments have to be incredibly well attuned to public attitudes. How can they channel some of this knowledge and ongoing engagement with the public towards more open and fundamental questions about science, technology and innovation?

Can we be too precautionary?

Critics might argue that upstream public engagement brings with it the danger of being too cautious. If R&D processes are opened up to too much public scrutiny, will innovation be stifled, preventing the emergence of new technologies? Should companies be left free to innovate in order to achieve much-needed technological break-throughs?

Where you stand in this debate depends on your views of the 'precautionary principle' – a widely used, and even more widely misinterpreted, concept that forms the cornerstone of European environmental and health policy. The most commonly-used definition of the precautionary principle is the one agreed at the 1992 Rio Earth Summit: 'where there are threats of serious or irreversible damage, lack of full scientific certainty shall not be used as a reason for postponing cost-effective measures to prevent environmental degradation.'[81] In other words, the fact that there is scientific uncertainty should not be used as an excuse to do nothing. Though this principle is widely quoted, there is little consensus about what it means in practice, and even less agreement about how strictly it should be applied.

The organisation *Spiked!* recently conducted a survey of scientists, asking them which historic scientific achievements would have been thwarted by applying the precautionary principle. The answers, ranging from vaccinations and the contraceptive pill through to the internet and iron, made a powerful case for the pitfalls of precaution.[82] However, the opposite question also needs to be asked – what tragedies could have been prevented through early and effective use of the precautionary principle? A study by the European Environment Agency points to a range of problems, from collapsing fish stocks to asbestosis, which could have been prevented if a precautionary approach had been taken. The hazards of working with asbestos were first documented as early as 1898, yet it took another hundred years – and millions of deaths – before it was finally banned in the EU. Between 1898 and 1998, evidence gradually accumulated of the harmful effects of asbestos, yet regulators were slow to respond. A more precautionary approach would, in this instance, have saved lives.[83]

Faced with these different viewpoints, it can seem hard to decide whether we invoke the precautionary principle too often, or too rarely. It is tempting to see this as a question of what balance to strike between precaution and innovation – to what extent scientists and companies should be free to experiment, or when and how to rein them in. But the trade-off between precaution and innovation may not be this straightforward. There are a couple of reasons why a more precautionary approach may actually stimulate, rather than stifle, innovation.

First, evidence suggests that carefully-designed regulation can promote innovation by encouraging leading companies to try different approaches. The European Environment Agency uses the example of asbestos: 'tighter regulation of asbestos would have raised its market price . . . thereby stimulating the innovation that belatedly led to better and often cheaper substitutes, as well as to improved engine and building designs that generate, at source, less waste heat.'[84] Similarly, Japanese manufacturers of fuel-efficient cars and German packaging companies have both benefited from tight regulations which provided a domestic market and, subsequently, export opportunities when standards in other countries caught up.

Second, a precautionary approach that engages the public may provide a way of anticipating potential problems before products reach the marketplace. Take GM foods as an example. Here, the new technology was not properly introduced to the public until products hit the supermarket shelves – at which point, the backlash began. A more precautionary approach would have allowed companies to understand and incorporate views and values, preventing a boom-and-bust cycle of innovation.

The precautionary principle will only stimulate innovation in this way if it is used wisely. It should not be crudely interpreted along the lines of 'if it can't be proved safe, it shouldn't be allowed'. And it should not mean taking the least risk option in all cases. NGOs are often too quick to invoke the precautionary principle to justify a ban, which can be unhelpful. It is better that we see precaution, and the public's involvement in framing and interpreting it, as just one more process through which complex decisions can be made – a process which may help to bring us a few steps closer to see-through science.

5. See-through science

I believe the intellectual life of the whole of western society is increasingly being split into two polar groups . . . at one pole we have the literary intellectuals . . . at the other scientists. . . . Between the two a gulf of mutual incomprehension – sometimes (particularly among the young) hostility and dislike, but most of all lack of understanding. They have a curious distorted image of each other.

CP Snow[85]

If he had lived to meet him, one wonders what CP Snow would make of Rob Doubleday. Snow, who achieved success as both a scientist and a novelist, is best remembered for his 1959 lecture *The Two Cultures*, in which he lamented the breakdown in communication between the sciences and the humanities. Doubleday, on the other hand, is a social scientist who recently took a job in the nanoscience laboratory at Cambridge University, providing real-time reflection on the social and ethical aspects of its research. 'My role', explains Doubleday, 'is to help imagine what the social dimensions might be, even though the eventual applications of the science aren't yet clear.' Communication is a big part of his work: 'A lot of what I do is translate and facilitate conversations between nanoscientists and social scientists, but also with NGOs and civil society.'[86]

This is a different sort of experiment to the experiments in democracy that we discussed in chapter 3. But it is no less important. Taking public engagement upstream requires us to be creative in the mix of formal and informal methods that are used to democratise science and infuse it with new forms of public knowledge.

Taken to its logical conclusions, our argument in this pamphlet has profound implications for the future of science. At its most ambitious, can upstream engagement reshape not only the way that science relates to public decision-making, but also the very foundations of knowledge on which the scientific enterprise rests? Five years on from the House of Lords report, this is the question that the science and society agenda now needs to address.

Running through our analysis is the proposition that different types of intelligence need to viewed alongside one another, rather than in a hierarchy which places science above the public. Why? Because this will lead to better science. Better in instrumental terms, because if scientists engage as equals in a dialogue with the public at an early stage, the likelihood of clashes further downstream is reduced. But also better in substantive terms: science that embraces these plural and diverse forms of knowledge will be more socially-robust science. As Helga Nowotny puts it, 'Science can and will become enriched by taking in the social knowledge it needs in order to continue its stupendous efficiency in enlarging our understanding of the world.'[87]

Paddling upstream

As we have seen, the science community has travelled a long way in a short time. In less than 20 years, the style of its conversation with society has changed from the patronising tones of 'public understanding' to the warmer banter of dialogue. Now it is changing again, to a more honest and reflective mode of listening and exchange.

Welcome to see-through science.

Will the rhetorical commitments to upstream engagement made by the government and the Royal Society be backed up by meaningful actions? It is still too early to say. The doubling of the budget for science and society activities in the new ten-year strategy is certainly welcome, and will help to resource new efforts and alliances.[88] But money alone cannot bring about the transformation we have described. This agenda requires many organisations and individuals to commit to institutional and cultural change. So as we head upstream, what are the landmarks along the way? Where are the sites for action and intervention?

Science policy

The priority for government is to deliver on the commitment to upstream public engagement contained in the ten-year science strategy. The practical and institutional mechanisms for this need to be worked through, and a strategic framework put in place. As the Royal Society report suggests, nanotechnologies are an obvious focal point for experimenting with new methods and approaches. We outline below what a *Nano Nation?* process might look like, drawing on our earlier discussion of the criteria for successful engagement.

Another eye-catching recommendation in the Royal Society report is for the chief scientific adviser to 'establish a group that brings together representatives of a wide range of stakeholders to look at new and emerging technologies and identify at the earliest possible stage areas where potential health, safety, environmental, social, ethical and regulatory issues may arise and advise on how these might be addressed.'[89] This is an excellent proposal which we would support, perhaps with the small addition that any such group should include lay members as well as representatives of interest groups.

Nano Nation?
Drawing on the lessons from *GM Nation?*, how should a public debate on nanotechnologies be run?

1. Clarity about outcomes
The government should be transparent from the start about how the outcomes of *Nano Nation?* would be used in its decision-

making process. One of the main shortcomings of *GM Nation?* was that this was never clear. It is not a case of just agreeing to abide by the outcomes of the debate – rather, the government and other participating institutions need to explain up front how they intend to use the process to improve their 'social intelligence' about nanotechnologies, in order to reach decisions that take account of public views and values.

2. Genuine deliberation
Nano Nation? should be a deliberative process. It should allow people to form and revise their views in discussion with others. This means that a range of methods and activities will be required:

o   public engagement techniques, including deliberative mapping and citizens' juries, which allow experts and the public to exchange views;

o   targeted processes for particular groups, including scientists, social scientists, economists, environment and development NGOs;

o   links to wider civil society, for example by encouraging newspapers to involve their readers in the debate, or asking companies with an interest in nanotechnology to involve their consumers;

o   links to the political process, by asking MPs to debate with constituents, allowing time for debate in Parliament, and ensuring that all ministers (and not just the science minister) discuss the issue;

o   a sufficient time period to enable learning and reflection to take place across all of these different activities. Another criticism of *GM Nation?* was that it was conducted in isolation from other elements of the GM policy process.

3. Debate informing research

Nano Nation? should set the agenda for further research on the social, ethical and environmental dimensions of nanotechnology. The debate process should be used to inform research priorities, rather than government, scientists or other experts deciding what questions should be answered. The Royal Society has made clear what further research *it* believes is necessary – but this may be different from public concerns. In particular, it is important that the process is not unevenly tilted towards narrow framings of risk if these do not accurately reflect public concerns. This is not to say that public views should be privileged over expert views – rather, that the input of expert and public knowledge should inform the way the debate proceeds.

4. Virtuous learning circles

Once *Nano Nation?* has taken place, its results must be revisited as developments across different nanotechnologies gather pace. Smaller, reconvened dialogues involving people engaged in the original debate could be used to revisit issues and help frame future research. When new regulations for nanotechnologies are drawn up, it should be clear how these have taken account of the public debate.

5. From local and national to European and beyond

The findings of *Nano Nation?* should inform the UK's stance in EU and international debates. The UK would then be able to stress that its position on nanotechnologies is thoroughly grounded in public views. This is important in international forums, such as the World Trade Organization, which currently place far more emphasis on scientific evidence. The UK should also argue for deliberative processes to be embedded in international regulations and decision-making.

Research councils

The research councils have a potentially decisive role in determining whether upstream public engagement becomes a meaningful reality. Across the councils, there have been encouraging moves to involve the public and other stakeholders in recent years, through the use of advisory panels and consultation exercises. Yet progress across the different councils is patchy. Some, such as BBSRC and the Medical Research Council, are making significant progress. Others appear to be lagging behind. The reform of research council structures now needs to move up a gear. Academic scientists and industrialists still monopolise the top layers of governance within the councils – the boards and central committees that set overall funding priorities and strategic direction. These need to be opened up to more diverse forms of social expertise and public knowledge.

Similar reforms are needed at intermediary levels within the councils, such as the selection and framing of research programmes and thematic priorities. While it may not be practical to have lay members on every single funding panel, neither is it acceptable to rule out wider public scrutiny of funding decisions on the basis that they are too technical. More urgency needs to be applied to the task of identifying what models of engagement and participation are most appropriate in particular contexts.

Related to this are questions of how funds are allocated. For example, in the US, research into the 'societal and educational implications' of nanotechnology are built into all funding programmes at the National Nanotechnology Initiative (NNI). Around 4 per cent of their 2003 budget was allocated to these activities.[90] The councils should explore the allocation of research funds in key areas to upstream public engagement and other forms of social evaluation. Careful consideration would need to be given to how this money was best spent – some might be directed towards interdisciplinary partnerships between natural/physical scientists and social scientists.

A related suggestion made by the biologist Rupert Sheldrake is for a '1 per cent fund' – a small percentage of the overall science budget that could be spent in carrying out research in areas that the public feel are important or neglected by other funding.[91] Participative methodologies could be used to determine areas of popular choice.

The practice of science

Rob Doubleday's work in the Cambridge nanoscience lab highlights another area where different forms of intelligence and public knowledge can be integrated into the everyday processes of doing science. Other labs should explore whether they can follow the Cambridge example by appointing social scientists to carry out what Dave Guston and Dan Sarewitz call 'real-time technology assessment'.[92]

No one is suggesting that we invite members of the public to stand over the shoulder of scientists while they work in the laboratory, but it is important to identify ways in which processes of engagement can strengthen the *reflective capacity* of scientists. In part, this is about 'bringing out the public' in the scientist – scientists are parents, children and citizens like everyone else, and are of course quite capable of reflecting on social and ethical questions. However, they may also benefit from occasional contact with non-expert and lay perspectives on the broader social implications of their work. A change is required in the scientific mindset, away from assumptions that experimentation begins and ends in the laboratory, and towards a recognition that experimentation continues as scientific and technological knowledge diffuses into complex social systems.[93]

New partnerships

We close with recommendations aimed at three sectors that will have important contributions to make to a culture of see-through science:

Companies

In line with our discussion in chapter 4, R&D-based companies should open up their innovation processes at the earliest possible stage, to ensure that a broader set of social insights are brought to bear on developments. Particular resources for this process may be found in management theories of open innovation and in processes for corporate sustainability and CSR.

NGOs

NGOs often taken a strong campaigning stance against particular technologies, for example Greenpeace's fight against GM, or the ETC Group's campaigns on nanotechnology. In a world of upstream engagement, NGOs will, themselves, need to move upstream, by talking to scientists, businesses and policy-makers at an early stage, and making their views and concerns clear from the start. The subtle and intelligent stance that Greenpeace has adopted towards nanotechnology perhaps offers a model that others could follow.

The media

Journalists are often caricatured by the science community as the source of public misinformation and concern. Reports on science and society often end with a call for more science or science specialists in the media to improve the public understanding of science. We have no intention of doing likewise. As Ian Hargreaves and colleagues point out 'a "science for science's sake" approach seems the one least likely to generate public engagement'. Instead, the idea of the *public interest* is central to engaging the public in science stories. 'We need to ask what it is important for citizens to know about science in a democracy. . . . What matters here, we would suggest, is not so much the science itself, but establishing clear connections between science, policy and the broader public interest.'[94]

Notes

1   S Hilgartner, *Science on Stage: expert advice as public drama* (Stanford: Stanford University Press, 2000), p 6.

2   The Royal Society and Royal Academy of Engineering, *Nanoscience and Nanotechnologies: opportunities and uncertainties* (London: The Royal Society, July 2004).

3   The Royal Society, *Risk: analysis, perception and management* (London: The Royal Society, 1992).

4   Telephone interview, 2 August 2004.

5   B Page, 'Public attitudes to science', *Renewal* 12, no 2 (London: Lawrence & Wishart, 2004).

6   S Jasanoff, *Designs on Nature* (Princeton University Press, forthcoming).

7   B Wynne, 'The public understanding of science' in S Jasanoff, GE Markle, JC Peterson and T Pinch (eds), *Handbook of Science and Technology Studies* (Thousand Oaks, CA: Sage, 1995), pp 361–88.

8   The Royal Society, *The Public Understanding of Science* (London: The Royal Society, 1985).

9   House of Lords Select Committee on Science and Technology, *Science and Society* (London: House of Lords, 23 February 2000).

10 Ibid, paragraph 3.9. This quote comes from Sir (now Lord) Robert May's evidence to the committee.

11 Ibid, paragraph 5.1.

12 See, for example, L Wolpert, 'Expertise required', *Times Higher Education Supplement*, 19 Dec 2003; D Taverne, 'How science can save the world's poor', *Guardian*, 3 Mar 2004.

13 R Willis and J Wilsdon, 'Technology, risk and the environment' in A Giddens (ed.), *The Progressive Manifesto* (Cambridge: Polity Press, 2003).

14 The Royal Society and Royal Academy of Engineering, *Nanoscience and Nanotechnologies*, p 64.

15 Ibid, p xi.

16   DTI, 'Nanotechnology offers potential to bring jobs, investment and prosperity – Lord Sainsbury', Department of Trade and Industry, press release, 29 July 2004.

17   HM Treasury/DTI/DfES, *Science and Innovation Investment Framework 2004–2014* (London: HM Treasury, July 2004), p 105.

18   For example, see Ibid, p 105.

19   Jasanoff, *Designs on Nature*.

20   HM Treasury/DTI/DfES, *Science and Innovation Investment Framework 2004–2014*, p 1.

21   T Blair, *Science Matters*, speech to the Royal Society, 23 May 2002.

22   A Stirling, 'Opening Up or Closing Down? Analysis, participation and power in the social appraisal of technology' in M Leach, I Scoones and B Wynne (eds), *Science, Citizenship and Globalization* (London: Zed Books, forthcoming).

23   Jasanoff, *Designs on Nature*.

24   DH Guston and D Sarewitz, 'Real-time technology assessment', *Technology in Society* 24 no 1 (2002): 93–109.

25   A Rip, T Misa and J Schot (eds), *Managing Technology in Society: the approach of constructive technology assessment* (London: Thomson, 1995).

26   J Durant, 'An experiment in democracy' in S Joss and J Durant (eds), *Public Participation in Science: the role of consensus conferences in Europe* (London: Science Museum, 1995).

27   Conversation with Robin Grove-White, 21 May 2004.

28   Following Sheila Jasanoff's work on the centrality of political culture in shaping public and policy responses to developments in science and technology, we are cautious about claiming that these arguments can easily be applied outside of the UK without greater adaptation to local contexts.

29   Remarks at a NATO press briefing, Brussels, 6 August 2002.

30   J Ezard, 'Rumsfeld's unknown unknowns take prize', *Guardian*, 2 Dec 2003.

31   For example, R Grove-White, P Macnaghten and B Wynne, *Wising Up: the public and new technologies* (Lancaster: CSEC/IEPPP, 2000); C Marris, B Wynne, P Simmons and S Weldon, *Public Perceptions of Agricultural Biotechnologies in Europe* (May 2002); available at: www.lancs.ac.uk/depts/ieppp/pabe/; GM Nation? Public Debate Steering Board, *GM Nation? The findings of the public debate*, available at: www.gmnation.org.uk.

32   Grove-White, Macnaghten and Wynne, *Wising Up*, p 29.

33   M Power, *The Risk Management of Everything: rethinking the politics of uncertainty* (London: Demos, 2004).

34   Ibid, p 11.

35   Strategy Unit, *Risk: improving government's capability to handle risk and uncertainty* (London: Cabinet Office, November 2002).

36   U Beck, *Risk Society: towards a new modernity* (London: Sage, 1992).

37   B Wynne, 'Risk as globalizing "democratic" discourse? Framing subjects and citizens' in M Leach, I Scoones and B Wynne (eds), *Science, Citizenship and Globalization* (London: Zed Books, forthcoming).

38   Ibid.

39   For a good overview of this work, see J Petts, J Homan and S Pollard, *Participatory Risk Assessment: involving lay audiences in environmental decisions on risk; R&D Technical Report E2-043/TR/01* (Bristol: Environment Agency, 2003).

40   Z Bauman, *Modernity and Ambivalence* (Cambridge: Polity Press, 1991).

41   Jasanoff, *Designs on Nature*.

42   Wynne, 'Risk as globalizing "democratic" discourse? Framing subjects and citizens'.

43   B Cooke and U Kothari (eds), *Participation: the new tyranny?* (London: Zed Books, 2001).

44   Wynne, 'Risk as globalizing "democratic" discourse? Framing subjects and citizens'.

45   M Midgley, *Science as Salvation: a modern myth and its meaning* (London: Routledge, 1992), p 15.

46   SustainAbility, *Who Needs It? Market implications of sustainable lifestyles* (London: SustainAbility, 1995).

47   C Leadbeater, *Mind over Matter: greening the new economy* (London: Green Alliance, 2000).

48   V Vaitheeswaran, *Power to the People: how the coming energy revolution will transform an industry, change our lives, and maybe even save the planet* (New York: Farrar, Straus and Giroux, 2003), p 25.

49   For more on this point, see Green Alliance's *Energy Entrepreneurs* project, available at: www.green-alliance.org.uk.

50   EF Schumacher, *Small is Beautiful* (London: Vintage, 1973), p 128.

51   Ibid, p 127.

52   The Royal Society and Royal Academy of Engineering, *Nanoscience and Nanotechnologies*, p 10.

53   R Grove-White, M Kearnes, P Miller, P Macnaghten, J Wilsdon and B Wynne, *Bio-to-Nano? Learning the lessons, interrogating the comparison* (Lancaster University/Demos, June 2004), p 9.

54   House of Commons Science and Technology Committee, *Too little too late? Government investment in nanotechnology* (London: The Stationery Office, 2004).

55   For a fuller account of these different categories, see J Wilsdon 'The politics of small things: nanoscience, risk and democracy' in *Renewal* 12, no 2, (London: Lawrence & Wishart, 2004).

56   KE Drexler, *Engines of Creation: the coming era of nanotechnology* (New York: Random House, 1986).

57   National Science Foundation, MC Roco and WS Bainbridge (eds), *Converging Technologies for Improving Human Performance* (Arlington, VA: NSF/DOC, June 2002).

58   See, for example, The Expectations Network, available at: www.york.ac.uk/org/satsu/expectations/index.htm; M Borup and K Konrad, 'Expectations in nanotechnology and in energy – foresight in the sea of

expectations', paper for a Research Workshop on Expectations in Science and Technology, 29–30 April 2004, Risø, Denmark; N Brown and M Michael, 'A sociology of expectations: retrospecting prospects and prospecting retrospects', *Technology Analysis & Strategic Management* 15, no 1 (2003): 4–18.

59   University of East Anglia Understanding Risk programme, *A Deliberative Future? An independent evaluation of the GM Nation? public debate about the possible commericalisation of transgenic crops in Britain, 2003* (Norwich: University of East Anglia, 2004).

60   The Royal Society and Royal Academy of Engineering, *Nanoscience and Nanotechnologies*, p 82.

61   National Consumer Council, *Winning the Risk Game* (London: National Consumer Council, 2003), p 21.

62   Stirling, 'Opening Up or Closing Down?'.

63   Ibid.

64   Ibid.

65   This list is adapted from POST, *Open Channels: public dialogue in science and technology* (London: Parliamentary Office of Science and Technology, 2001).

66   See www.deliberative-mapping.org.uk for further details.

67   Jasanoff, *Designs on Nature*.

68   See, for example, J Dryzek, *Deliberative Democracy and Beyond: liberals, critics, contestations* (Oxford: Oxford University Press, 2000); and F Fischer, *Citizens, Experts and the Environment: the politics of local knowledge* (Durham and London: Duke University Press, 2000).

69   GM Nation? Steering Board, *GM Nation? The findings of the public debate*, available at: www.gmnation.org.uk.

70   The Electoral Commission and the Hansard Society, *An Audit of Political Engagement: Research Report 2004* (London: The Electoral Commission and the Hansard Society, 2004) summary p 3.

71   RSA Forum for Technology, Citizens and the Market, *What's there to talk about? Public engagement by science-based companies in the UK* (London: RSA, June 2004), p 21.

72   HM Treasury/DTI/DfES, *Science and Innovation Investment Framework 2004–2014*, p 55.

73   Grove-White, Macnaghten and Wynne, *Wising Up*. For a broader summary of this issue, see S Krimsky, *Science in the Private Interest: has the lure of profits corrupted biomedical research?* (Oxford: Rowman and Littlefield, 2003).

74   See www.btplc.com/Societyandenvironment/Hottopics/Hottopics.htm.

75   R Grove-White, P Macnaghten, S Mayer and B Wynne, *Uncertain World: genetically modified organisms, food and public attitudes in Britain* (Lancaster/CSEC, 1997).

76   T Macmillan, *Engaging in Innovation: towards an integrated science policy* (London: IPPR, July 2004).

77   HM Treasury/DTI/DfES, *Science and Innovation Investment Framework 2004–2014*, p 71.

78   See, for example, H Chesbrough, *Open Innovation: the new imperative for*

creating and profiting from technology (Boston, MA: Harvard Business School Press, 2003).

79   C Leadbeater, *Open Innovation: the European model*, research proposal, 2003.

80   R Jupp, *Getting Down to Business: an agenda for corporate social innovation* (London: Demos, 2001).

81   United Nations, *Rio Declaration on Environment and Development*, 1992, available at: www.unep.org/Documents/Default.asp?DocumentID=78&ArticleID=1163.

82   S Starr, 'Science, risk and the price of precaution', available at: www.spiked-online.com, 1 May 2003.

83   European Environment Agency, *Late Lessons from Early Warnings: the precautionary principle 1896–2000* (Copenhagen: European Environment Agency, 2001).

84   Ibid, p 59.

85   CP Snow, *The Two Cultures* (Cambridge: CUP, 1959).

86   Interview, 25 July 2004.

87   H Nowotny, P Scott and M Gibbons, *Rethinking Science: knowledge and the public in an age of uncertainty* (Cambridge: Polity Press, 2001), p 262.

88   The science and society budget is set to increase from £4.25 million per year in 2005/06 to over £9 million per year by 2006/07.

89   The Royal Society and Royal Academy of Engineering, *Nanoscience and Nanotechnologies*, p 87.

90   Market Research Society Bulletin June 2003, available at: www.mrs.org.uk/publications/bulletin.

91   R Sheldrake, 'Set them free', *New Scientist* 19 Apr 2003.

92   Guston and Sarewitz, 'Real-time technology assessment'.

93   A point made by Sheila Jasanoff in a speech at the CESAGen conference, 2 March 2004.

94   I Hargreaves, J Lewis and T Spears, *Towards a Better Map: science, the public and the media* (Swindon: ESRC, 2003).

DEMOS – Licence to Publish

compensation. The exchange of the Work for other copyrighted works by means of digital file-sharing or otherwise shall not be considered to be intended for or directed toward commercial advantage or private monetary compensation, provided there is no payment of any monetary compensation in connection with the exchange of copyrighted works.

c If you distribute, publicly display, publicly perform, or publicly digitally perform the Work or any Collective Works, You must keep intact all copyright notices for the Work and give the Original Author credit reasonable to the medium or means You are utilizing by conveying the name (or pseudonym if applicable) of the Original Author if supplied; the title of the Work if supplied. Such credit may be implemented in any reasonable manner; provided, however, that in the case of a Collective Work, at a minimum such credit will appear where any other comparable authorship credit appears and in a manner at least as prominent as such other comparable authorship credit.

5. Representations, Warranties and Disclaimer
a By offering the Work for public release under this Licence, Licensor represents and warrants that, to the best of Licensor's knowledge after reasonable inquiry:
  i Licensor has secured all rights in the Work necessary to grant the licence rights hereunder and to permit the lawful exercise of the rights granted hereunder without You having any obligation to pay any royalties, compulsory licence fees, residuals or any other payments;
  ii The Work does not infringe the copyright, trademark, publicity rights, common law rights or any other right of any third party or constitute defamation, invasion of privacy or other tortious injury to any third party.
b EXCEPT AS EXPRESSLY STATED IN THIS LICENCE OR OTHERWISE AGREED IN WRITING OR REQUIRED BY APPLICABLE LAW, THE WORK IS LICENCED ON AN "AS IS" BASIS, WITHOUT WARRANTIES OF ANY KIND, EITHER EXPRESS OR IMPLIED INCLUDING, WITHOUT LIMITATION, ANY WARRANTIES REGARDING THE CONTENTS OR ACCURACY OF THE WORK.

6. Limitation on Liability. EXCEPT TO THE EXTENT REQUIRED BY APPLICABLE LAW, AND EXCEPT FOR DAMAGES ARISING FROM LIABILITY TO A THIRD PARTY RESULTING FROM BREACH OF THE WARRANTIES IN SECTION 5, IN NO EVENT WILL LICENSOR BE LIABLE TO YOU ON ANY LEGAL THEORY FOR ANY SPECIAL, INCIDENTAL, CONSEQUENTIAL, PUNITIVE OR EXEMPLARY DAMAGES ARISING OUT OF THIS LICENCE OR THE USE OF THE WORK, EVEN IF LICENSOR HAS BEEN ADVISED OF THE POSSIBILITY OF SUCH DAMAGES.

7. Termination
a This Licence and the rights granted hereunder will terminate automatically upon any breach by You of the terms of this Licence. Individuals or entities who have received Collective Works from You under this Licence, however, will not have their licences terminated provided such individuals or entities remain in full compliance with those licences. Sections 1, 2, 5, 6, 7, and 8 will survive any termination of this Licence.
b Subject to the above terms and conditions, the licence granted here is perpetual (for the duration of the applicable copyright in the Work). Notwithstanding the above, Licensor reserves the right to release the Work under different licence terms or to stop distributing the Work at any time; provided, however that any such election will not serve to withdraw this Licence (or any other licence that has been, or is required to be, granted under the terms of this Licence), and this Licence will continue in full force and effect unless terminated as stated above.

8. Miscellaneous
a Each time You distribute or publicly digitally perform the Work or a Collective Work, DEMOS offers to the recipient a licence to the Work on the same terms and conditions as the licence granted to You under this Licence.
b If any provision of this Licence is invalid or unenforceable under applicable law, it shall not affect the validity or enforceability of the remainder of the terms of this Licence, and without further action by the parties to this agreement, such provision shall be reformed to the minimum extent necessary to make such provision valid and enforceable.
c No term or provision of this Licence shall be deemed waived and no breach consented to unless such waiver or consent shall be in writing and signed by the party to be charged with such waiver or consent.
d This Licence constitutes the entire agreement between the parties with respect to the Work licensed here. There are no understandings, agreements or representations with respect to the Work not specified here. Licensor shall not be bound by any additional provisions that may appear in any communication from You. This Licence may not be modified without the mutual written agreement of DEMOS and You.